K. CONNORS

Curious Chronicles

A Journey Through Uncommon Knowledge and Strange Facts

Copyright © 2024 by K. Connors

All rights reserved. No part of this publication may be reproduced, stored or transmitted in any form or by any means, electronic, mechanical, photocopying, recording, scanning, or otherwise without written permission from the publisher. It is illegal to copy this book, post it to a website, or distribute it by any other means without permission.

First edition

This book was professionally typeset on Reedsy.
Find out more at reedsy.com

Contents

Introduction	1
Chapter 1: The Hidden Histories	3
Chapter 2: The Wonders of Nature	8
Chapter 3: Anomalies of the Mind and Body	14
Chapter 4: Mysteries of the Deep	20
Chapter 5: Technological Oddities	32
Chapter 6: Lost and Found	38
Chapter 7: Cultural Curiosities	44
Chapter 8: Paranormal and Unexplained	51
Chapter 9: Pioneers of the Unusual	57
Conclusion	63

Introduction

Hey there, fellow seeker of the strange and purveyor of the peculiar! Welcome to "Curious Chronicles: A Journey Through Uncommon Knowledge and Strange Facts," your unofficial guidebook to the wonderfully weird and wackily wonderful nooks and crannies of our world. This isn't your standard encyclopedia or a dry dissertation on the drab and dreary. Oh no, my friend. This is a roller coaster ride through the odd, the bizarre, and the downright hilarious aspects of human history, nature, technology, and those things that go bump in the night (or in the case of some insects, things that go "buzz" in your ear at 3 a.m.).

Suppose you've ever wondered about the kind of stuff that doesn't make it into mainstream textbooks, like why people in Gloucestershire hurl themselves down a steep hill in pursuit of a rolling cheese, or how a certain inventor decided that what the world really needed was a mechanical device to tip one's hat. In that case, you, my curious compadre, have stumbled upon your literary soulmate.

Within these pages, you'll find tales so bizarre they'll make you question the fabric of reality. We'll explore ancient technologies that have modern scientists scratching their heads, delve into historical oddities that prove truth is stranger than fiction, and meet individuals whose eccentricities left an indelible mark on the world (or at least made for some really interesting dinner party conversation).

But wait, there's more! We'll also venture into the great outdoors, not just to hug trees (though that's encouraged) but to marvel at nature's penchant for the dramatic and the unexpected. From murderous plants to animals with identity crises, Mother Nature proves time and again that she's got a wicked sense of humor.

And because no exploration of the odd and unusual would be complete without a foray into the paranormal, we'll light our torches and tiptoe into the shadowy realms of ghosts, cryptids, and UFOs. Whether you're a die-hard believer in the supernatural or the kind of person who reads ghost stories with a skeptical eyebrow raised, there's something here to tickle your fancy or fuel your nightmares (you're welcome).

So, why embark on this journey through the atlas of the absurd? Because, dear reader, in a world that often takes itself too seriously, it's a breath of fresh, albeit slightly bizarre, air to remember that curiosity, wonder, and a healthy dose of laughter are essential to the human experience. This book is a celebration of the oddballs, the eccentrics, and the trailblazers who remind us that being normal is overrated and that the most interesting stories are often found off the beaten path.

Prepare to be amused, confused, and enthused as we dive headfirst into the curious chronicles of our weird and wacky world. Buckle up, it's going to be a wonderfully weird ride!

Chapter 1: The Hidden Histories

Welcome to the shadowy corridors of history, where the light of common knowledge seldom reaches. Here, in the nooks and crannies of the past, lie stories so bizarre, so unexpected, that they seem more like the product of a fevered imagination than the cold, hard facts of history. But fear not, intrepid reader, for we are about to embark on a journey to uncover these hidden histories, guided by the dim torch of curiosity and the slightly unreliable compass of humor.

The Forgotten Inventor of... Sliced Bread?

Let's start with something so mundane, so utterly commonplace, that you probably never gave it a second thought: sliced bread. Yes, the greatest thing before... well, sliced bread. But who do we have to thank for this pinnacle of convenience? Otto Frederick Rohwedder, that's who. Otto, an inventor who clearly understood the plight of uneven toast, toiled away in obscurity to bring us the bread-slicing machine in 1928. Legend has it that when he first proposed his idea, people thought he was a few crumbs short of a loaf. Today, we stand on the shoulders of this giant, wielding our knives only at will, not necessity.

The Pirate Who Defied Death... and Accounting

Next up, let's hoist the Jolly Roger and set sail with the most feared (and perhaps the most financially savvy) pirate of the Caribbean, "Black Bart" Roberts. Roberts was not only notorious for capturing over 400 ships but also for his strict code of conduct and democratic principles. He was an early adopter of workers' rights, ensuring his crew received fair shares of the loot and health insurance benefits. Yes, you heard right—pirate health insurance! If a pirate lost a limb in the line of duty, he was compensated with a hefty sum. It makes you wonder if corporate boardrooms today could learn a thing or two from Black Bart's pirate governance.

The Emperor of the United States, Protector of Mexico

History books often neglect to mention that the United States once had an emperor. No, not George Washington or Abraham Lincoln, but Emperor Norton I, self-declared ruler of the United States and protector of Mexico. In 1859, Joshua A. Norton, a failed businessman turned self-proclaimed monarch, wandered the streets of San Francisco issuing decrees, abolishing Congress, and demanding the construction of a bridge connecting San Francisco to Oakland. The remarkable thing? People loved him for it. Businesses accepted his self-made currency, and he dined at the finest restaurants without charge. Emperor Norton's legacy is a testament to the power of charisma, eccentricity, and, perhaps, the collective need for a bit of madness in our lives.

When the Eiffel Tower Was Nearly a Scrap Heap

The Eiffel Tower, that iron lattice symbol of romance, was originally intended as a temporary installation for the 1889 World's Fair. Gustave Eiffel, the engineer behind the tower, faced immense criticism from Paris's artistic community, who deemed it a monstrosity. They couldn't

CHAPTER 1: THE HIDDEN HISTORIES

wait for it to be dismantled. Yet, here we stand, over a century later, with the tower still piercing the Parisian skyline. Its salvation? The tower's unexpected utility as a giant radio antenna. That's right, the Eiffel Tower was too useful to tear down, proving that sometimes utility trumps beauty, even in the heart of Paris.

The Great Emu War: Birds 1, Australian Army 0

To wrap up our tour of history's hidden corridors, let's travel to Australia, where in 1932, the Australian Army waged war against... emus. Yes, the flightless birds. Following World War I, veterans were given land in Western Australia for farming, but they soon found themselves competing with a burgeoning emu population for crops. The solution? Military intervention. Armed with machine guns, the Australian Army set out to tackle the emu problem, only to find their foe surprisingly resilient. The emus scattered in the face of gunfire, making them difficult targets. After weeks of fruitless effort, the military withdrew, and the emus claimed victory. The Great Emu War remains a humorous anecdote of man's attempt to control nature and failing spectacularly.

Continuing our jaunt through the annals of the overlooked and underappreciated, let's delve deeper into the quirky corners of history that textbooks dare not tread. These stories not only add flavor to our understanding of the past but remind us that truth can indeed be stranger—and more entertaining—than fiction.

The Day the Sun Didn't Rise

Imagine waking up to find that the sun has decided to take a day off. This is not the setup for a science fiction story but an actual event that occurred on May 19, 1780, known as New England's Dark Day. The residents of the northeastern United States woke to complete darkness at noon, with no explanation in sight. Candles were lit, animals were confused, and many believed the apocalypse had arrived. The cause? A combination of smoke from forest fires, thick fog, and cloud cover. Yet, in the absence of a clear explanation, fear and speculation ran rampant. It's a stark reminder of how dependent we are on the natural world and how unsettling it can be when it behaves unexpectedly.

The Man Who Sold the Eiffel Tower. Twice.

Meet Victor Lustig, a con artist so audacious he sold the Eiffel Tower. Not once, but twice. In the 1920s, Lustig posed as a government official and convinced a group of scrap metal dealers that the Eiffel Tower was being dismantled and that he was authorized to sell it for scrap. One dealer, eager for the deal of a lifetime, paid a hefty sum before realizing he'd been duped. Lustig vanished with the money, only to pull the same stunt a second time. His escapades serve as a testament to the power of persuasion and the allure of a too-good-to-be-true deal. Lustig's legacy is a cautionary tale about the fine line between genius and infamy.

The Battle of the Bucket

Wars have been fought over land, over religion, and over power, but did you know one was fought over a bucket? The War of the Bucket, or the Battle of Zappolino in 1325, was waged between the rival city-states of Bologna and Modena in Italy. The conflict ignited over the theft of a ceremonial bucket from a well in Bologna by Modenese raiders. The ensuing battle saw thousands of soldiers clashing, with Modena emerging victorious—and the bucket remaining in Modena to this day. It's a stark reminder of how seemingly trivial disputes can escalate into serious conflicts, a lesson that remains relevant in our own times.

CHAPTER 1: THE HIDDEN HISTORIES

The First Animal in Space Wasn't a Monkey

While many believe the first animal in space was a monkey, it was actually a dog named Laika, sent into orbit by the Soviet Union in 1957. Laika, a stray from the streets of Moscow, was chosen for her calm demeanor and small size. Her journey aboard Sputnik 2 marked a significant, albeit controversial, milestone in space exploration. Though Laika did not survive the mission, her legacy lives on, highlighting the sacrifices made in the name of progress and sparking a global conversation on animal rights and ethics in scientific research.

The Mysterious Disappearance of the Ninth Legion

The Roman Empire's Ninth Legion, once a formidable force in Britain, vanished from history in the 2nd century AD. Theories about their disappearance range from defeat by Scottish tribes to being transferred to the Middle East. Despite extensive historical and archaeological investigations, the fate of the Ninth Legion remains one of history's enduring mysteries. Their disappearance is a haunting reminder of the impermanence of even the mightiest armies and the limits of our historical knowledge.

These stories, from the darkened skies of New England to the audacious scams of Victor Lustig, from ancient wars over buckets to space-faring dogs, illustrate the breadth and depth of human experience. History is more than a collection of dates and events; it's a tapestry woven from the incredible, the improbable, and the downright bizarre. As we close this chapter on hidden histories, let's carry forward a sense of wonder and a keen eye for the stories lurking in the shadows, waiting for their moment in the sun.

Chapter 2: The Wonders of Nature

Welcome back, fellow adventurers, to our ongoing expedition through the wild and woolly realms of uncommon knowledge and strange facts. If Chapter 1 had us time-traveling through history's hidden alleyways, Chapter 2 plants us firmly on the lush, unpredictable grounds of Mother Nature's estate. Here, the dress code is strictly bizarre, and the guest list is as varied as it is eccentric. So, strap on your hiking boots, apply a generous layer of curiosity, and let's delve into the wonders of nature that prove reality often outstrips the imagination.

The Animal Kingdom's Oddballs

First on our nature trail is the animal kingdom, home to creatures so bizarre you'd think they were dreamed up during a particularly wild brainstorming session between Mother Nature and Salvador Dalí. Take the platypus, for example, an animal so odd that when European scientists first encountered a specimen, they thought it was a hoax. With the bill of a duck, the tail of a beaver, and the feet of an otter, this mammal lays eggs but also produces milk, making it a walking contradiction and a prime example of nature's flair for improvisation.

Not to be outdone, the narwhal, often dubbed the 'unicorn of the sea', sports a massive tooth that protrudes from its head like a jousting lance. Scientists believe this tusk is used in mating rituals to impress potential partners, proving that in the animal kingdom, style points do count.

Vegetation with a Vendetta

Moving from animals to plants, the flora of our world is no less peculiar or fascinating. Consider the Venus flytrap, a carnivorous plant with a taste for insects. This botanical oddity lures its prey with nectar, then snaps shut faster than a mousetrap, digesting its unsuspecting victim with enzymes. It's the plant kingdom's version of a horror movie, with a botanical twist.

Then there's the "walking" palm tree of the Amazon rainforest, known as Socratea exorrhiza. Unlike its stationary cousins, this tree can literally move towards the light by growing new roots in the desired direction and letting the old ones die off. It's the botanical equivalent of picking up your house and moving to a sunnier spot because you decided you needed more natural light.

Weather Wonders

If you thought weather was just sunshine and rain clouds, think again. Nature has a few meteorological tricks up her sleeve that could easily feature in a fantasy novel. Take ball lightning, for example, a rare phenomenon where lightning forms into a sphere and can float or move erratically before exploding. Reports of ball lightning have baffled scientists for centuries, with some encounters even suggesting these glowing orbs can pass through walls. It's like nature's own version of a ghost story, but with a scientific twist.

Then there's "blood rain," a weather event that sounds like it was plucked from the pages of a Gothic horror novel. This phenomenon occurs when rain mixes with sand from deserts, turning the water a deep, unsettling red. It's been reported from ancient times to the present day, turning skies and landscapes into scenes that look more like Mars than Earth.

The Uncharted Depths

The ocean covers more than 70% of our planet's surface, yet we've explored less of it than we have outer space. Within its uncharted depths lurk creatures that defy the imagination. The giant squid, for example, long considered a sailor's tall tale, has been proven to exist, with tentacles capable of ensnaring a whale. These elusive creatures can grow up to 43 feet long, embodying the very essence of maritime mystery and adventure.

Not to be outshone, the blobfish, a gelatinous mass dwelling in the deep sea, was once voted the world's ugliest animal. With a face only a mother could love, the blobfish's comical appearance belies its adaptability to extreme pressure, making it a marvel of evolutionary ingenuity.

Geological Jigsaw Puzzles

Turning our gaze to the ground beneath our feet, the Earth's geological features offer their own set of curiosities. Take the sailing stones of Death Valley, for instance. These rocks, some weighing hundreds of pounds, mysteriously move across the desert floor, leaving trails in their wake. The explanation? A perfect storm of ice, water, and wind conditions. It's as if the Earth is playing its own slow-motion game of chess.

CHAPTER 2: THE WONDERS OF NATURE

Then there's the Door to Hell, a natural gas field in Turkmenistan that's been burning continuously since it was lit by scientists in 1971. Originally intended to prevent the spread of methane gas, the crater has become a fiery spectacle, a literal glimpse into what the entrance of the underworld might look like if it were managed by the Department of Natural Resources.

Continuing our exploration of nature's cabinet of curiosities, let's delve even deeper into the treasure trove of oddities that our planet offers. These tales of the natural world serve as a reminder that, despite our advancements in science and technology, there are still mysteries out there that elude explanation, tickling our sense of wonder and challenging our understanding of the ordinary.

The Musical Sands of the Desert

Imagine walking across a desert and hearing the sands beneath your feet sing. This isn't a scene from a fantasy novel but a real phenomenon occurring in several deserts around the world. The "singing sands," as they are often called, emit a musical note when the sand grains slide over each other under the right conditions. Scientists believe the shape and silica content of the sand grains play a crucial role in this natural concert. It's as if the earth itself is trying to communicate through its own version of a string instrument, proving that nature might just be the original musician.

The Everlasting Lightning Storm

In Venezuela, at the mouth of the Catatumbo River, there's a storm that seems to defy the very laws of nature. The Catatumbo lightning occurs for up to 160 nights a year, 10 hours per night, and up to 280 times per hour. This never-ending storm has served as a natural lighthouse for sailors and has even entered the Guinness World Records for the highest concentration of lightning. The phenomenon is caused by a mix of topography and wind patterns, creating a natural spectacle that illuminates the night sky with a relentless display of nature's power. It's as though Zeus himself decided to set up a permanent residence in Venezuela.

The Boiling River of the Amazon

Deep in the Peruvian Amazon lies a river so hot that it boils. The Shanay-Timpishka, known as the "boiling river," reaches temperatures up to 212 degrees Fahrenheit. This natural hot spring, stretching for nearly 4 miles, is not heated by volcanic activity, as one might expect, but by geothermal heat from the Earth's interior. The river is both a sacred site for local indigenous peoples and a natural wonder that challenges our expectations of where and how water can exist in liquid form. It's a stark reminder that even water, the most familiar of substances, can hold surprises in the right conditions.

Dancing Forests and Crooked Trees

In a small corner of Kaliningrad, Russia, lies the Curonian Spit, home to the "Dancing Forest," a peculiar grove of pine trees that twist and turn in bizarre shapes and loops. The cause of this arboreal ballet is still debated, with theories ranging from human intervention to natural phenomena. Similarly, the "Crooked Forest" in Poland features a group of pine trees with a 90-degree bend at their base, forming a curious and unnatural-looking forest. These forests challenge our perceptions of growth and gravity, proving that even trees can march to the beat of their own drum.

CHAPTER 2: THE WONDERS OF NATURE

The Underwater Waterfall Illusion

Off the coast of Mauritius, an incredible natural illusion creates the appearance of an underwater waterfall. This visual wonder is actually a trick of the light and the geography of the ocean floor. Sand and silt deposits cascade down the ocean shelf, creating the illusion of a waterfall plunging into the depths. From above, the scene is breathtaking, a testament to the beauty and mystery that lies in the interplay of light, water, and earth. It's a vivid reminder that sometimes, nature's most astonishing wonders are not what they seem at first glance.

These stories, from singing sands and perpetual storms to boiling rivers and enigmatic forests, invite us to look closer at the world around us, to question the familiar, and to marvel at the mysteries that remain unsolved. Nature, in all its diversity and complexity, continues to be a source of inspiration, wonder, and, occasionally, amusement. So let us keep exploring, with open eyes and minds, ready to be surprised by the next chapter of natural wonders waiting just around the corner.

Chapter 3: Anomalies of the Mind and Body

Buckle up, dear readers, for we are about to embark on a journey into the bewildering, awe-inspiring realm of human anomalies. This chapter is not for the faint of heart, for we will delve into the mysteries of the mind and body that defy explanation, challenge our understanding of biology, and occasionally make us squirm in our seats. From individuals with extraordinary abilities that seem lifted from the pages of comic books to those with medical conditions so rare they've baffled science, this exploration celebrates the diversity and resilience of the human species.

Michel Lotito: The Man Who Ate an Airplane

First up is Michel Lotito, a man whose diet makes the adventurous eaters among us look positively conservative. Known as "Monsieur Mangetout" (Mr. Eat-All), Lotito consumed metal, glass, and rubber, among other materials. Over his lifetime, he ate an entire Cessna 150 airplane, a feat that took him two years to complete. His extraordinary ability was due to a condition known as pica, a disorder characterized by an appetite for non-nutritive substances, combined with a thick lining in his stomach and intestines that allowed him to consume sharp metal without injury. Lotito reminds us that the phrase "you are what you eat" has never been more complex.

CHAPTER 3: ANOMALIES OF THE MIND AND BODY

The Woman Who Sees 100 Million Colors

Imagine seeing the world with 100 times the color perception of the average person. This is the reality for the rare "tetrachromats," individuals with an extra type of cone cell in their eyes. One such person, an artist by necessity, perceives colors most of us couldn't dream of, turning a mundane trip to the grocery store into a psychedelic experience. This extraordinary condition challenges our understanding of reality, proving that the way we see the world is as unique as our DNA.

Wim Hof: The Iceman Cometh

Wim Hof, known as "The Iceman," possesses the seemingly superhuman ability to withstand freezing temperatures that would send the rest of us scrambling for the nearest blanket. Hof attributes his ability to control his body temperature in extreme cold to a combination of meditation, breathing techniques, and exposure training. He has run a marathon above the Arctic Circle in shorts, swum under ice for extended distances, and stood covered in ice cubes for nearly two hours. Hof's mind-over-matter approach has not only earned him world records but also invites us to reconsider the limits of human endurance and the power of the mind.

Stephen Wiltshire: The Human Camera

Stephen Wiltshire, an artist with an extraordinary photographic memory, can draw detailed landscapes and cityscapes from memory after just a brief observation. Diagnosed with autism at a young age, Wiltshire's talent showcases the incredible ways in which the human brain can process and recall information. His detailed drawings of cities, down to the exact number of windows in skyscrapers, remind us that the mind can unlock capabilities beyond our wildest imaginations.

The Mystery of Synesthesia

Enter the world of synesthesia, a condition where the stimulation of one sensory or cognitive pathway leads to automatic, involuntary experiences in a second sensory or cognitive pathway. Some synesthetes can hear colors, see sounds, or taste shapes, blending the senses into a unique experience of the world. This neurological phenomenon challenges our perceptions of reality and illustrates the vast spectrum of human sensory experience. It serves as a vivid reminder that not everyone experiences the world in the same way, and what we consider 'normal' is merely a matter of perspective.

Daniel Kish: Seeing with Sound

Daniel Kish, blind since he was a baby, uses echolocation to navigate the world around him. By making clicking sounds with his mouth and listening to the echoes that bounce back from objects, he can identify their size, shape, and location. Kish's ability, often associated with bats and dolphins, has earned him the nickname "the real-life Batman." He teaches this technique to other visually impaired individuals through his nonprofit organization, challenging our assumptions about the limitations of human senses and the adaptability of the human spirit.

As we continue to unravel the tapestry of human anomalies, our journey reveals not just extremes of physical and mental capabilities, but also the profound adaptability and resilience that define the human condition. These additional stories highlight the remarkable ways individuals have harnessed their unique conditions to change perceptions, break barriers, and inspire those around them.

CHAPTER 3: ANOMALIES OF THE MIND AND BODY

The Real-life Superheroes of Memory

There are individuals among us who possess what can only be described as superhuman memory. These memory athletes, as they're often called, can recall an astonishing array of information, from the order of multiple decks of cards after a single viewing to thousands of digits of pi. While some of this ability is innate, much is also honed through techniques and training, challenging the myth that we are limited by our natural capacities. Their feats prompt us to reconsider the untapped potential of our own minds and the power of dedicated mental exercise.

The Girl Who Feels No Pain

Imagine a life without pain. For most, it sounds like a dream, but for those with congenital insensitivity to pain, it's a dangerous reality. This rare condition means they can endure cuts, burns, and injuries without feeling discomfort, leading to a high risk of unnoticed harm. However, individuals with this condition, like a young girl named Gabby, navigate life with extraordinary caution and bravery, teaching us about the complexity of pain as both a curse and a crucial, protective sensation. Their stories are a powerful reminder of the body's innate warning system and the importance of safeguarding against harm even when the alarm bells are silent.

The Yogis of Extreme Meditation

In the serene yet extreme world of meditation, some practitioners have taken their mental and physical discipline to astonishing lengths. These modern-day yogis demonstrate the incredible control one can have over the body's responses to stress, pain, and temperature. Documented cases include slowing one's heart rate, altering body temperature, and even surviving without food or water for periods that defy scientific explanation. This profound mastery over the mind-body connection underscores the untapped power of human concentration and willpower, suggesting that the key to unlocking extraordinary physical feats lies in the depths of mental discipline.

The Color-changing Eyes of Elizabeth Taylor

Hollywood legend Elizabeth Taylor was famous for many things, including her acting prowess, tumultuous love life, and striking beauty. One of her most mesmerizing features was her violet eyes, a trait so rare it contributed to her allure and mystique. However, Taylor's eye color was not a simple genetic quirk but a combination of natural pigmentation and the way her eyes reflected light, sometimes appearing to change color under different lighting conditions. Her captivating gaze reminds us that beauty and anomaly often walk hand in hand, challenging our perceptions of normalcy and the enchanting variety found within human genetics.

The Man Who Can Run Forever

Lastly, we turn to the extraordinary tale of Dean Karnazes, an ultramarathon runner capable of feats that would exhaust even the most seasoned athletes. Karnazes has run 50 marathons in 50 states in 50 consecutive days, a 350-mile race without sleep for over 80 hours, and a marathon to the South Pole in sub-zero temperatures. His body's unique ability to rapidly clear lactate from his muscles allows him to endure long distances without the fatigue that limits most humans. Karnazes's story is a testament to the remarkable adaptability of the human body and the spirit of endurance that drives some individuals to push the boundaries of possibility.

CHAPTER 3: ANOMALIES OF THE MIND AND BODY

These narratives of extraordinary human anomalies serve as powerful reminders of our species' diversity, resilience, and the boundless potential that lies within each of us. They encourage us to look beyond the ordinary, to celebrate the unique qualities that differentiate us, and to pursue the limits of our own capabilities with curiosity and courage. As we close this chapter, let us carry forward the inspiration to explore the mysteries of the mind and body, uncovering the hidden talents and abilities that reside within us all.

Given the collaborative and sequential nature of this book project and considering the word count and content delivery preferences you've outlined, let's proceed to the next chapter with an engaging narrative and a splash of humor to keep things lively.

Chapter 4: Mysteries of the Deep

Dive in, folks, as we submerge into the enigmatic world beneath the waves in this chapter of "Curious Chronicles." The ocean, that vast, mysterious blue that hugs our continents, holds secrets so bizarre and wonders so profound, they make the tales of Atlantis seem like a children's bedtime story. From underwater rivers to singing sand, the ocean's mysteries are as deep as they are captivating.

The Underwater Rivers of the Black Sea

Let's start with a phenomenon that sounds straight out of a fantasy novel: underwater rivers. In the depths of the Black Sea, scientists have discovered rivers, complete with banks and floodplains, flowing beneath the surface. These subaquatic rivers are formed by salty water from the Mediterranean, which is denser than the Black Sea's fresh water, creating a separate, flowing layer. It's a mesmerizing sight, akin to flying over a terrestrial river valley, if you could fly and breathe underwater, that is. This discovery challenges our perceptions of what a river can be and where it can exist, proving once again that water always finds a way.

The Singing Sands of the Caribbean

CHAPTER 4: MYSTERIES OF THE DEEP

Next, imagine standing on a beach in the Caribbean, not just any beach, but one where the sand beneath your feet sings. Yes, you read that right—singing sand. This rare phenomenon occurs when the grains are perfectly round and of similar size, creating a squeaking or singing sound when walked upon. The exact mechanics of this concert are still under study, but it's thought that the sound is generated by the friction between the sand grains. It's nature's own choir, and the only admission ticket you need is your presence... and maybe a gentle foot shuffle.

The Immortal Jellyfish: Nature's Benjamin Button

Dive deeper with me now, to meet Turritopsis dohrnii, the jellyfish that laughs in the face of Father Time. This tiny creature has the remarkable ability to revert back to its juvenile polyp stage after reaching maturity, essentially living its life in reverse. It's the closest thing nature has to a real-life Fountain of Youth, and it's swimming around in our oceans right now. The implications of this discovery are profound, offering insights into cellular regeneration and aging. Plus, it's a great party fact to share when the conversation takes a dive into the morbid inevitability of aging.

The Lost City of Heracleion

For those who thought lost cities were only the stuff of legend, let's visit Heracleion. Once a bustling metropolis of the ancient world, it vanished into the depths of the Mediterranean Sea. Rediscovered in 2000 near the coast of Egypt, Heracleion is a time capsule of temples, statues, and artifacts. Its discovery is akin to finding Atlantis, except it's real, and archaeologists are the treasure hunters mapping its streets and recovering its stories. The city's sinking remains a topic of study, but it's believed to have been caused by a combination of subsidence, earthquakes, and rising sea levels. It's a humbling reminder of nature's power and the impermanence of even the greatest human achievements.

The Great Blue Hole: A Portal to the Deep

Off the coast of Belize lies the Great Blue Hole, a giant marine sinkhole that looks like Earth's own pupil when viewed from above. This underwater cavern descends into the dark abyss for about 400 feet, its walls teeming with life and geological history. Originally a limestone cave formed during the last ice age, it was flooded and transformed into the submarine wonder we see today. Divers who have ventured into its depths describe it as a journey to another world, where stalactites and ancient formations tell the story of Earth's climatic shifts. It's a stark reminder of the planet's age and the fleeting nature of human history in comparison.

The Dancing Yet Deadly Brinicles

Lastly, let's chill with the brinicles, also known as "icicles of death." These icy fingers of doom form under sea ice when a flow of extremely cold, saline water meets the warmer ocean water. As they grow, reaching toward the seafloor, they create a web of ice that freezes everything in its path, including unfortunate sea creatures. Witnessing a brinicle in action is like watching a slow-motion dance of destruction, beautiful yet deadly. It's a fascinating example of the dynamic and sometimes harsh interactions within Earth's ecosystems.

Continuing our dive into the ocean's mysterious depths, let's explore even more phenomena that blur the line between scientific reality and the kind of tales you'd expect to find in a fantasy novel. The ocean, with its uncharted territories and alien-like inhabitants, never ceases to amaze, offering endless stories that captivate the curious mind.

The Bioluminescent Bays of Vieques

Venture with me now to the magical shores of Vieques, Puerto Rico, where the waters glow with an ethereal light. The Mosquito Bay is home to one of the most spectacular displays of bioluminescence in the world, courtesy of tiny dinoflagellates that light up when disturbed. Paddling through these waters at night is like stirring the stars into the sea, creating a galaxy beneath your fingertips. This natural light show isn't just stunning; it's a glowing testament to the beauty and complexity of life on Earth. Scientists study these glowing creatures to understand more about bioluminescence, but for those lucky enough to see it in person, it's like being part of a living dream.

The Mysterious Milford Sound Underwater Forest

New Zealand's Milford Sound, a fjord renowned for its breathtaking landscapes, hides a submerged secret: an underwater forest. Divers can swim among the trees of this drowned forest, which remains preserved in the cold, dark waters. The sight is surreal, as if nature decided to flip the script and grow a forest in the depths instead of on land. The trees, victims of past landslides, create an eerie, otherworldly habitat for marine life. Exploring this aquatic woodland offers a rare glimpse into a unique ecosystem where the line between terrestrial and marine becomes beautifully blurred.

The Thermal Vents: Nature's Underwater Factories

Deep on the ocean floor, far from the reach of sunlight, lie hydrothermal vents, spewing superheated water rich in minerals. These vents form bizarre landscapes that could rival any alien planet's surface. They're not just geological wonders; they're biological hotspots where life thrives in extreme conditions. Around these vents, communities of giant tube worms, clams, and unique microbial life forms have adapted to harness the chemical energy released, proving that life can flourish in the most inhospitable places. These vents challenge our understanding of life's limits and hint at the possibilities for life beyond Earth.

The Salty Mirror: The Bonneville Salt Flats Underwater

Utah's Bonneville Salt Flats are known for land speed records, but during the wet season, they transform into a vast, shallow lake, creating a perfect mirror that reflects the sky. This stunning phenomenon blurs the horizon line, making it difficult to tell where the earth ends and the sky begins. For photographers and dreamers alike, the salt flats offer a glimpse into a world without borders, where the beauty of the sky is echoed at your feet. This natural spectacle reminds us of the transformative power of water, even in the most arid landscapes.

The Vanishing Lake of Mount Gambier

Australia's Blue Lake in Mount Gambier is known for its dramatic seasonal color transformation. For most of the year, the lake's waters are a cool steel grey, but in November, like clockwork, it turns a vibrant cobalt blue, a change that mystified observers for years. Recent studies suggest the color shift is due to the warming waters, which change the angle of sunlight penetration and affect the way light is scattered and absorbed. This natural phenomenon is a stunning reminder of the dynamic nature of our planet's ecosystems and the visual magic that can occur when conditions align perfectly.

CHAPTER 4: MYSTERIES OF THE DEEP

As we conclude this chapter on the mysteries of the deep, it's evident that our oceans are a source of endless fascination and wonder. From glowing bays and submerged forests to alien-like thermal vents and transformative lakes, the depths of our seas hold stories that challenge our imagination and expand our understanding of the natural world. These tales of the deep serve as a reminder of the planet's beauty, complexity, and the continuous interplay between the elements. So, as we look out across the vast expanse of water that covers our Earth, let's remember that there are still countless mysteries waiting to be discovered, beneath the waves and beyond.

Chapter 5: Cosmic Conundrums

Blast off with us as we journey beyond the blue skies and into the star-studded expanse of the universe. Space, the final frontier, is filled with mysteries that have puzzled humans since the dawn of time. From planets that rain glass sideways to the enigmatic dark matter holding galaxies together, the cosmos is like a cosmic jigsaw puzzle with most of the pieces missing. So, strap in and prepare for a tour of the universe's most bewildering cosmic conundrums.

The Planet with Iron Rain

Imagine a world so hot that metals vaporize into the atmosphere, only to rain down as molten iron. Welcome to WASP-76b, an exoplanet where temperatures reach 4,400 degrees Celsius on the side facing its star, causing metals like iron to evaporate into the air. As strong winds carry these vapors to the cooler night side, the iron condenses into droplets, creating a downpour of iron rain. This bizarre weather pattern is a stark reminder of the wild diversity of planetary climates out there, far beyond our tame Earthly weather systems.

The Mystery of Dark Matter

Dark matter, the universe's greatest mystery, is the invisible scaffolding that holds galaxies together. Despite making up about 85% of the universe's mass, it doesn't emit, absorb, or reflect light, making it completely invisible and detectable only by its gravitational effects. Scientists are still scratching their heads over what dark matter could be, with theories ranging from weakly interacting massive particles (WIMPs) to axions. This cosmic enigma challenges our understanding of the universe's very fabric and is a central puzzle in modern astrophysics.

The Galactic Cannibals

In the cosmic dance of gravity, not all galaxies play nice. Some, known as galactic cannibals, grow by devouring their smaller neighbors. The Milky Way itself is on a collision course with the Andromeda Galaxy, with an expected meet-up in about 4 billion years. This slow-motion cosmic feast is a dramatic reminder of the scale and dynamism of the universe, where even galaxies are not immune to the forces of change and evolution.

The Boomerang Nebula: The Coldest Place in the Universe

Forget winter in Siberia; the coldest place in the universe is the Boomerang Nebula, where temperatures dip to a chilling -272°C, just a degree above absolute zero. This pre-planetary nebula, located about 5,000 light-years away, achieves its frigid temperatures through the rapid expansion of gas ejected by a dying star. It's a cosmic freezer that out-chills the vast void of space itself, providing a unique laboratory for studying the physics of extreme cold.

CHAPTER 4: MYSTERIES OF THE DEEP

The Pulsar with the Precision of an Atomic Clock

Pulsars, the spinning remnants of supernovae, are cosmic lighthouses that emit regular bursts of radiation with the precision of atomic clocks. One in particular, PSR J1748-2446ad, spins at a mind-boggling 716 times per second, making it the universe's fastest-spinning known object. These stellar remnants are not just astronomical curiosities; they're also key tools for testing the limits of physics, including the theory of relativity and the search for gravitational waves.

The Planet of Diamond

55 Cancri e, an exoplanet twice the size of Earth, is believed to be composed largely of carbon in the form of diamond. With a third of its mass thought to be pure diamond, this planet offers a glimpse into the staggering variety of planetary compositions possible in the universe. The discovery challenges our Earth-centric view of planetary formation and composition, proving that the universe has more in its treasure chest than we could have ever imagined.

The Eternal Darkness of Rogue Planets

Rogue planets, celestial orphans without a star to call home, wander the galaxy in eternal darkness. These nomadic worlds, ejected from their solar systems by gravitational disturbances, drift through the cosmos, cold and alone. The existence of rogue planets challenges our traditional notions of planetary systems and hints at the vast, unexplored diversity of objects in our galaxy.

The Unpredictable Quasars

Quasars, the brightest objects in the universe, are powered by supermassive black holes at the centers of distant galaxies. Emitting more light than an entire galaxy, these cosmic beacons can outshine everything else in the universe. Their light, traveling billions of years to reach us, offers a glimpse into the universe's early days. However, the mechanisms of their incredible luminosity and the ways they impact their host galaxies remain among the most compelling mysteries in astrophysics.

The Enigma of Fast Radio Bursts

Fast Radio Bursts (FRBs), brief flashes of radio waves from deep space, have puzzled scientists since their discovery. Lasting only milliseconds, these bursts pack a punch, emitting as much energy as the sun does in a day. The sources and causes of FRBs are still under debate, with theories ranging from neutron stars to extraterrestrial intelligence. These cosmic signals underscore the universe's vastness and the abundance of phenomena yet to be understood.

As we further navigate the cosmic sea, let's delve into additional enigmas that underscore the universe's penchant for the peculiar and profound. Each cosmic phenomenon we encounter not only broadens our understanding of the cosmos but also highlights the incredible journey science is on, unraveling the mysteries of existence.

The Ever-Expanding Universe

One of the most fundamental and awe-inspiring revelations of modern astronomy is the discovery that the universe is expanding at an accelerating rate. The force behind this acceleration remains one of the greatest mysteries in physics, attributed to an enigmatic energy known as dark energy. This mysterious force, making up about 68%

of the universe, works in opposition to gravity and is responsible for the increasing pace at which galaxies are moving apart. This expansion challenges our notions of the cosmos and raises profound questions about the universe's ultimate fate. Will it expand forever, or is there a limit to its growth? The answers lie in understanding the true nature of dark energy, a puzzle that continues to baffle the brightest minds.

The Ghostly Neutrinos

Neutrinos, often called "ghost particles," are nearly massless particles that barely interact with matter, making them incredibly difficult to detect. Trillions of neutrinos pass through our bodies every second, yet we remain oblivious to their presence. These elusive particles are believed to hold critical clues to several cosmic mysteries, including the workings of the sun and the dynamics of supernova explosions. The study of neutrinos challenges our understanding of particle physics and offers a unique window into processes at the heart of stars and galaxies.

The Strange Case of the Missing Antimatter

The universe presents us with a perplexing imbalance: the mysterious absence of antimatter. Theoretical physics suggests that the Big Bang should have produced equal amounts of matter and antimatter, yet our observable universe is overwhelmingly composed of matter. What happened to the antimatter? Various hypotheses have been proposed, from asymmetries in the laws of physics to the annihilation of matter and antimatter in the early universe. This cosmic discrepancy is a crucial puzzle in understanding the fundamental forces and conditions of the early universe.

The Enigmatic Supermassive Black Holes

At the heart of nearly every galaxy, including our own Milky Way, lurks a supermassive black hole, millions to billions of times the mass of the sun. The formation of these gravitational behemoths remains one of the greatest enigmas in astronomy. How did they grow so massive? Did they form from the collapse of giant gas clouds, or from the mergers of smaller black holes? The lifecycle of supermassive black holes and their influence on galaxy formation and evolution are key to deciphering the narrative of the cosmos.

The Cosmic Web

The universe at its largest scales is structured like an intricate web, with galaxies and galaxy clusters connected by filaments of dark matter, separated by vast voids. This cosmic web is a direct consequence of the universe's initial conditions and the dark matter that shapes its large-scale structure. The distribution and dynamics of this web, observed through the gravitational lensing of light around these massive structures, offer insights into the universe's very fabric. Yet, understanding the full extent of this web and its implications for cosmic evolution remains a daunting challenge for cosmologists.

These cosmic conundrums, from the accelerating expansion of the universe to the ethereal dance of neutrinos, the puzzling shortage of antimatter, the mysteries surrounding supermassive black holes, and the grand architecture of the cosmic web, highlight the profound complexity and beauty of the universe. Each mystery invites us into a deeper exploration of the cosmos, encouraging us to ponder, research, and marvel at the wonders beyond our world. As we continue this celestial journey, let us embrace the unknown with curiosity and humility, reminded that the universe is far more vast and enigmatic than we could have ever imagined. The quest for knowledge is endless, and the

CHAPTER 4: MYSTERIES OF THE DEEP

cosmos, with its infinite mysteries, is our ultimate frontier, beckoning us to uncover its secrets and perhaps, in the process, learn more about ourselves.

Chapter 5: Technological Oddities

Welcome to the wild, wacky world of technological oddities, where innovation meets the "what in the world were they thinking?" From gadgets that make you scratch your head to inventions that seem to defy logic, this chapter explores the curious corners of human ingenuity. So, buckle up as we take a tour through the annals of technological history, showcasing inventions that are as bizarre as they are brilliant.

The Useless Box: A Masterpiece of Futility

First up, let's give it up for the Useless Box, a device that truly lives up to its name. Flip the switch on, and the box springs to life, only to turn itself off again. That's it. That's the whole show. It's the embodiment of futility, yet there's something deeply satisfying about a machine that exists solely to cease its existence. Invented as a humorous critique of our obsession with gadgets, the Useless Box is a philosophical statement wrapped in a wooden box, proving that sometimes, the journey is more important than the destination—especially when the destination is nowhere.

The Pet Rock: The Ultimate Low-Maintenance Companion

CHAPTER 5: TECHNOLOGICAL ODDITIES

In the mid-1970s, the Pet Rock became a cultural phenomenon, making millions for its creator and leaving everyone else wondering why they didn't think of it first. It's exactly what it sounds like: a rock, in a box, with breathing holes (because, of course, rocks need air), and a comprehensive owner's manual. The Pet Rock was the ultimate low-maintenance pet, requiring no food, water, or walks, only a healthy suspension of disbelief. It's a testament to the power of marketing and perhaps a commentary on human loneliness. Or maybe it just proves that people will buy anything if it's packaged right.

The Flying Car: A Dream Deferred

The flying car is the quintessential symbol of future promise, featured in every second sci-fi movie and the dreams of commuters stuck in traffic. Various prototypes have taken to the skies over the years, from aerocars of the 1950s to modern VTOL (Vertical Take-Off and Landing) vehicles. Yet, the dream of a flying car in every driveway remains just that—a dream. Challenges range from safety and air traffic control to the simple fact that piloting a car is significantly harder than driving one. The flying car embodies our yearning for freedom and our penchant for overestimating our multitasking abilities. It's the technological equivalent of biting off more than you can chew, then trying to fly away with it.

The Smell-O-Vision: A Scent-sational Failure

Enter the Smell-O-Vision, a technology designed to bring movies to life by releasing relevant scents into the theater during the film. Introduced in the 1960s, it was meant to be the next big thing in cinema, adding olfactory experiences to the visual and auditory. However, timing issues and the overwhelming nature of certain scents led to its quick demise. Audiences were less than thrilled to have their movie experience interrupted by the sudden smell of flowers or, less appealingly, gunpowder. The Smell-O-Vision remains a quirky footnote in cinematic history, a reminder that not all of our senses are craving the limelight.

The Anti-Eating Face Mask: Dieting Through Dystopia

In the quest for weight loss solutions, the Anti-Eating Face Mask takes the cake (which, ironically, you wouldn't be able to eat if you were wearing it). Patented in 1982, this device was designed to prevent the wearer from indulging in solid food, promising a form of dieting that feels more like a horror movie prop. It's an extreme solution to overeating, highlighting the lengths to which people will go to conform to beauty standards, and a stark reminder that maybe, just maybe, there are better ways to address dietary habits than strapping on a face cage.

The Twitter-Peeping Toilets: Social Media's Final Frontier

In the age of oversharing, the Twitter-Peeping Toilets stand as a monument to the question, "But why?" These smart toilets analyze your deposits and tweet about your health (or lack thereof). While the intention might be to promote health awareness, one can't help but ponder the implications of your toilet joining the social media conversation. It's a bizarre blend of technology and privacy invasion, proving that just because we can connect everything to the internet doesn't necessarily mean we should.

The DVD Rewinder: A Solution in Search of a Problem

Ah, the DVD Rewinder, a product that flew off the shelves for reasons

CHAPTER 5: TECHNOLOGICAL ODDITIES

that remain a mystery to this day. In an era when DVDs were king, this device promised to "rewind" your DVDs, a task as unnecessary as trying to water your plastic plants. It's a shining example of a gadget that exists purely for the amusement of gifting it to the technologically clueless, a testament to human creativity's humorous side.

The Hushme: Privacy, but Make It Sci-Fi

In our ever-connected world, finding a private corner to make a phone call can be challenging. Enter the Hushme, a device that looks like something out of a sci-fi movie, designed to mask your voice during calls. Worn around the mouth, it ensures your conversations remain private, at the cost of making you look like a supervillain in training. It's a quirky solution to a modern problem, blending the desire for privacy with the undeniable appeal of looking like you're about to launch a space laser.

Diving deeper into the rabbit hole of technological marvels and misadventures, we continue our exploration with even more eyebrow-raising inventions that blur the line between genius and, well, the other thing. These gadgets, each more peculiar than the last, not only push the boundaries of what's possible but also question the boundaries of what's necessary. So, without further ado, let's proceed with our whimsical journey through the annals of odd tech.

The Banana Phone: Ring, Ring, Ring, Banana Phone!

In an era dominated by sleek, feature-packed smartphones, the Banana Phone stands out with its unapologetic simplicity and, dare we say, a-peel. This Bluetooth-enabled handset, shaped like a banana, connects to your smartphone and lets you take calls in a way that's sure to turn heads and raise questions about your produce preferences. While it might not offer the latest in mobile technology, it promises a bunch of laughs and curious glances. The Banana Phone is a reminder that sometimes, technology doesn't have to be serious to be seriously fun.

The USB Pet Rock: A Modern Twist on a Classic

Remember the Pet Rock from the 1970s? Well, it got a 21st-century update with the USB Pet Rock. This tech-infused stone comes with a USB cable that you can plug into your computer, and...that's it. It doesn't store data, it doesn't charge, and it certainly doesn't download. What it does offer is a steadfast companion that's immune to viruses, software updates, and just about anything else the digital world might throw at it. The USB Pet Rock is a testament to the enduring power of simplicity in an increasingly complex world.

The 360-Degree Selfie Toaster: Breakfast Meets Narcissism

In a world obsessed with selfies, the 360-Degree Selfie Toaster takes personalization to a whole new level. This invention allows you to burn your selfie onto your morning toast, ensuring that you can literally consume your own image with breakfast. While the practicality of eating your face might be debatable, it's a quirky way to start the day and a bold statement about our selfie-centric culture. It begs the question: if you eat a selfie-toast in the forest and no one is around to Instagram it, is it still delicious?

The Walking Sleeping Bag: Comfort on the Go

For those who've ever wished they could take the warmth of their bed wherever they go, the Walking Sleeping Bag is a dream come true. This

CHAPTER 5: TECHNOLOGICAL ODDITIES

invention combines the snug comfort of a sleeping bag with the mobility of, well, pants. Equipped with armholes and a zippered opening for your feet, this wearable sleeping bag ensures you're never without the cozy comforts of home, even when you're out in the wild. It's the perfect accessory for late-night snack runs or zombie apocalypses, proving that comfort doesn't have to be stationary.

The Submarine Sports Car: Because Why Not?

Last but certainly not least, we have the Submarine Sports Car, an invention that answers a question no one asked: what if you could drive your sports car underwater? This vehicle combines luxury sports car aesthetics with submarine functionality, allowing for underwater exploration at the push of a button. While its practical applications may be limited (and its price tag astronomical), it's a vivid example of engineering ingenuity and the human desire to conquer every terrain— land, sea, or anywhere your bank account allows.

As we conclude this chapter on technological oddities, it's clear that the spirit of invention knows no bounds. From the whimsically pointless to the surprisingly practical, these gadgets remind us of the thin line between genius and jest. They challenge us to rethink our definitions of necessity and innovation, all while providing a healthy dose of laughter. In a world that often takes itself too seriously, these technological oddities are a breath of fresh, albeit slightly bizarre, air. They encourage us to embrace our quirks, explore the edges of possibility, and remember that at the heart of every invention lies a simple desire: to create something that didn't exist before, no matter how outlandish it may seem.

Chapter 6: Lost and Found

Ah, the tantalizing tales of the lost and the suddenly found! This chapter takes a detour into the bizarre back alleys of history and geography, where we stumble upon the things humanity misplaced and then, sometimes by sheer luck, stumbled upon again. From entire cities that played hide and seek with civilization to valuable artifacts that decided to take an extended vacation from human eyes, these stories are a testament to our knack for losing and finding stuff in the most dramatic fashion.

The City That Took a Desert Sabbatical: Petra

Imagine misplacing something. Now, imagine misplacing an entire city. Welcome to Petra, the rose-red city half as old as time, which decided to take a little sabbatical in the Jordanian desert. Hidden away by towering rocks and a penchant for secrecy, Petra was lost to the Western world until 1812 when Swiss explorer Johann Ludwig Burckhardt decided to take the scenic route. Disguised as an Arab scholar, he stumbled upon the city that had been playing the world's most intense game of hide and seek. Petra's rediscovery reminds us that sometimes, you just need to take a wrong turn to make the right discovery.

CHAPTER 6: LOST AND FOUND

The Royal Library of Ashurbanipal: The OG Kindle

Long before digital libraries, there was the Royal Library of Ashurbanipal in Nineveh, modern-day Iraq. Boasting a collection of over 30,000 clay tablets covering everything from astrology to zoology, it was the Kindle of its day. Lost for over two millennia, it was rediscovered in the 19th century, buried under the sands of time, proving that even the ancients couldn't resist the charm of hoarding books they'd probably never get around to reading.

The Antikythera Mechanism: An Ancient Greek in a Modern World

The Antikythera Mechanism, often hailed as the world's first computer, was found by sponge divers off the coast of Greece in 1901. This intricate device, dating back to around 100 BC, was used to predict astronomical positions and eclipses. It's like finding an ancient smartphone in a world where everyone else is still marveling at the wheel. The discovery of the Antikythera Mechanism not only proved that the ancient Greeks were way ahead of their time but also that they probably wouldn't have been too impressed with our current tech.

The Lost Army of Cambyses: A Sand-Submerged Mystery

Around 524 BC, Cambyses II of Persia sent a 50,000-strong army into the Egyptian desert to attack the Oasis of Siwa. Legend has it that a massive sandstorm buried them all, and they were never seen again. Despite numerous attempts, no trace of the lost army has been found, making it one of history's greatest vanishing acts. It's as if the desert decided to sweep them under its colossal rug, reminding us that Mother Nature doesn't play favorites, especially with invading armies.

The S.S. Central America: The Ship of Gold

The S.S. Central America was a steamship that sank in a hurricane off the coast of the Carolinas in 1857, taking down with it a vast treasure of gold from the California Gold Rush. It was found over a century later, in 1988, turning the seabed into a high-stakes treasure map. The discovery of the "Ship of Gold" was a bonanza for treasure hunters and historians alike, proving that sometimes, X marks the spot for real.

The Rediscovery of Machu Picchu: Hiram Bingham's Accidental Tourist Attraction

When Hiram Bingham III "discovered" Machu Picchu in 1911 (because, apparently, the indigenous people who already knew about it didn't count), he was actually looking for a different lost city. Talk about a happy accident! Machu Picchu, perched high in the Andes, was one of the few major Incan cities to escape destruction by the Spanish, mainly because it was so well-hidden. Its rediscovery brought this architectural marvel into the limelight, turning it into a bucket list destination for tourists and llamas alike.

The Voynich Manuscript: The Book No One Can Read

Last but not least, let's talk about the Voynich Manuscript, a book that truly takes "I can't even" to new levels. Discovered in 1912 by a Polish book dealer, this illustrated manuscript is written in an unknown script, with bizarre illustrations that have puzzled linguists, cryptographers, and historians for over a century. It's the ultimate prank from the past—a book that no one can read, filled with pictures that make you wonder if medieval scribes were just trolling future generations.

CHAPTER 6: LOST AND FOUND

Continuing our treasure hunt through history's attic, let's dust off more tales of astonishing rediscoveries that remind us sometimes what's lost isn't gone forever, just taking a long nap. From ancient beverages resurrected from the depths of time to artworks hiding in plain sight, these stories are a testament to humanity's relentless pursuit of "Oh, so that's where I left it!"

The Brew That Time Forgot: Resurrecting Ancient Ales

Imagine finding a 5,000-year-old recipe for beer and thinking, "Let's brew this!" Well, some intrepid scientists and brewers did just that, using residue analysis from ancient pottery. These ancient ales offer a taste of history, quite literally, providing insights into our ancestors' sophisticated palates. It's like hosting a dinner party and serving a dish that's been out of fashion for a few millennia. Cheers to the ancients for their contribution to happy hour!

The Rediscovery of the Original "Star Wars" Model

In a galaxy not so far away (specifically, a storage unit in California), the original model of the Star Wars Millennium Falcon was found, years after its last appearance in the films. This piece of cinematic history had been gathering dust, forgotten, until it was rediscovered and restored to its former glory. It's a reminder that sometimes, the most valuable treasures are hiding in the back of our cosmic garage, waiting for their chance to shine again in the spotlight.

The Overlooked Masterpiece: Da Vinci's "Salvator Mundi"

"Salvator Mundi," attributed to Leonardo da Vinci, spent years masquerading as a mere copy before being recognized for the masterpiece it was. Purchased for a modest sum, it was later sold for over $450 million, making it one of the most expensive paintings ever sold. This story is akin to finding out the old painting in your attic is actually a long-lost Rembrandt, proving that sometimes, the art world's equivalent of a lottery win is just a cleaning and an expert eye away.

The Phoenix of the Baltic: The Amber Room

The Amber Room, an entire chamber decorated with amber panels backed with gold leaf and mirrors, was considered the eighth wonder of the world before it was lost during World War II. Despite extensive searches, its fate remains one of the greatest mysteries of modern times. The original might be lost, but a painstakingly crafted replica now stands in its place, a testament to the original's awe-inspiring beauty and the enduring allure of what's been lost to history.

Pompeii: The City Frozen in Time

The rediscovery of Pompeii in the 18th century opened a window into Roman life, frozen in time by the ash of Mount Vesuvius in AD 79. The city was lost for nearly 1,700 years, its existence reduced to a footnote in history until explorers unearthed its well-preserved streets. Walking through Pompeii is like time travel, offering an intimate glimpse into the daily lives of its inhabitants, from their homes and shops to their graffiti, proving that the Romans were not so different from us.

The Lost Colony of Roanoke: America's Enduring Mystery

The Lost Colony of Roanoke remains one of the greatest unsolved mysteries in American history. In 1587, over a hundred settlers vanished without a trace, leaving behind only the cryptic message "CROATOAN"

CHAPTER 6: LOST AND FOUND

carved into a tree. Despite numerous theories and exhaustive searches, their fate remains unknown. It's the historical equivalent of misplacing your keys, if your keys were an entire colony and you never found them again.

As we conclude our exploration of the world's most fascinating rediscoveries, it's clear that the past is never as far behind us as we might think. From ancient brews to cinematic icons, lost artworks to entire cities, these stories highlight the joy of rediscovery and the enduring human spirit of exploration. They remind us that history is not just a series of dates and events but a treasure trove of stories waiting to be uncovered, appreciated, and retold. So, keep your eyes peeled and your mind open; the next great discovery might just be under your nose, or perhaps, hiding in your attic.

Chapter 7: Cultural Curiosities

Pack your bags, don your explorer's hat, and ensure your sense of humor is securely fastened, for we're about to embark on a whirlwind tour of the world's most fascinating cultural curiosities. From traditions that make you go "Hmm..." to festivals that have you questioning the collective sanity of entire populations, this chapter is a testament to humanity's endless creativity in celebrating the weird and wonderful.

The Great British Cheese Rolling

Welcome to Gloucestershire, England, where the brave and the bold chase a 9-pound wheel of Double Gloucester cheese down a steep hill. The Cooper's Hill Cheese-Rolling and Wake is a tradition that defies common sense and safety regulations, drawing participants and spectators from around the globe. The goal? To catch the cheese, of course. The reality? A spectacular tumble down the hill, where the cheese invariably wins. It's a celebration of dairy and daring, proving that the British will put their necks on the line for a good piece of cheese.

Finland's Wife-Carrying Championship

In Finland, the key to marital bliss might just involve sprinting through

CHAPTER 7: CULTURAL CURIOSITIES

an obstacle course while carrying your spouse on your back. The Wife-Carrying Championship in Sonkajärvi is a sport that's as quirky as it is physically demanding. The prize? The wife's weight in beer. It's a tradition that raises important questions, like "Why?" and "But, seriously, why?" and yet, it's a beloved part of Finnish culture, celebrating strength, endurance, and the ability to not drop your spouse in a pool of mud.

Japan's Kanamara Matsuri: The Festival of the Steel Phallus

Japan's Kanamara Matsuri, or Festival of the Steel Phallus, takes openness about human anatomy to a whole new level. Celebrated annually in Kawasaki, the festival venerates fertility, marriage, and healthy childbirth with parades featuring giant phallic symbols. It's a striking example of cultural diversity, where reverence and humor walk hand in hand, and visitors are reminded that sometimes, it's okay to laugh with the gods.

The Spanish Tomatina: A Tomatoey Tussle

Imagine a food fight, but with thousands of participants and overripe tomatoes as the ammunition. Welcome to La Tomatina, Spain's messiest festival, held annually in the town of Buñol. For one hour, the streets turn red with tomato pulp, as revelers from around the world let loose in a squishy battle of epic proportions. It's a celebration that prompts us to reconsider the tomato, not just as a salad ingredient, but as a tool for joyous, albeit wasteful, abandon.

The Hair Freezing Contest of Yukon

In the frosty reaches of Yukon, Canada, there's a competition where the chillier your noggin, the better. The Hair Freezing Contest, held at the Takhini Hot Pools, invites participants to sculpt their wet hair into frozen works of art. The result? A gallery of icy coiffures that would make even the most avant-garde stylist blush. It's a testament to human creativity, resilience, and the questionable decisions we make for the sake of a good photo op.

The Scottish Highland Games: A Testament to Tartan Tenacity

The Scottish Highland Games combine the thrill of athletic competition with the pageantry of Scottish culture, featuring events like caber tossing (where participants hurl a large wooden pole) and the stone put (which is exactly what it sounds like). It's a celebration of strength, skill, and kilts, offering a glimpse into Scotland's rich history and the enduring spirit of its people. The Highland Games remind us that sometimes, throwing things and wearing skirts is not just fun, it's tradition.

The Quiet Village of Nagoro: Where Dolls Replace Departed Residents

In the remote village of Nagoro, Japan, dolls have taken the place of residents who've either moved away or passed on. Created by local artist Tsukimi Ayano, these life-sized dolls occupy abandoned homes, work fields, and school classrooms, creating a silent community that's both eerie and poignant. It's a unique response to rural depopulation, turning Nagoro into an open-air museum that explores themes of memory, loss, and the passage of time.

The Italian Battle of the Oranges

In the city of Ivrea, Italy, the historical reenactment of a medieval insurrection takes the form of the Battle of the Oranges. Teams engage in

CHAPTER 7: CULTURAL CURIOSITIES

a vigorous citrus exchange (read: hurl oranges at each other with gusto) to commemorate the city's defiance against tyranny. It's a vibrant, sticky affair that leaves the streets littered with orange pulp and participants nursing their vitamin C-infused bruises. The festival is a reminder of the power of community and the lengths to which people will go to remember their history, even if it means getting pelted by fruit.

As we continue our exploration of the globe's quirky customs and festivals, we dive deeper into the heart of human creativity and the sometimes baffling, always fascinating traditions that bind communities and cultures together. Let's keep our minds open and our giggles ready as we uncover more cultural curiosities that highlight the wonderfully weird world we share.

The Netherlands' Orange Madness: King's Day

In the Netherlands, King's Day (Koningsdag) turns the country into a sea of orange, celebrating the monarch's birthday with a fervor that could make a pumpkin patch jealous. The Dutch don everything orange, from clothes to cakes, in a national party that includes street markets, boat parades, and an unspoken competition for the most outrageously orange outfit. It's a day when the Dutch let loose, proving that when it comes to partying, they can give even the most exuberant cultures a run for their money. It's a reminder that sometimes, all you need for a good party is a reason to celebrate and a national obsession with a particular color.

The Curious Case of "Morning Glory" in Australia

In the remote town of Burketown, Australia, adventurers and glider pilots gather each spring to ride the "Morning Glory," a rare type of roll cloud. These long, tubular clouds can stretch up to 1,000 kilometers and move at speeds up to 60 kilometers per hour. Pilots soar on the dynamic air currents generated by these clouds, participating in a meteorological phenomenon that's as breathtaking as it is unique. It's an awe-inspiring example of nature's capacity to surprise and delight us, serving as a reminder that sometimes, the best adventures are those dictated by the whims of the weather.

The "Festival of the Exploding Hammer" in Mexico

In the Mexican state of San Luis Potosi, the "Festival of the Exploding Hammer" takes the idea of a bangin' party quite literally. Participants attach a mix of sulfur and chlorate to the ends of sledgehammers and then slam them against rail beams, creating a spectacular explosion. It's a tradition that combines the thrill of danger with the artistry of fireworks, in a way that makes health and safety officers everywhere reach for their stress balls. This explosive celebration is a testament to the human love for spectacle and the lengths to which people will go for a good show.

The Great Pantomime Horse Race of the UK

London's annual Great Pantomime Horse Race is where the noble steed meets slapstick comedy. Teams dressed in pantomime horse costumes race through the streets, navigating obstacles and stopping off at pubs, all in the name of charity. It's a spectacle that combines the competitiveness of a race with the absurdity of grown adults galloping in costume. This event showcases the British knack for combining humor with competition, proving that sometimes, the most memorable races aren't about speed but about the laughs along the way.

CHAPTER 7: CULTURAL CURIOSITIES

South Korea's Mud Festival

In Boryeong, South Korea, the Mud Festival celebrates the purported health benefits of mud with a week-long party of mud slides, mud wrestling, and even a mud king contest. Originating as a marketing stunt for cosmetics made from Boryeong mud, the festival has become a magnet for tourists and a testament to the fun of getting dirty. It's a celebration that encourages participants to cast aside adult inhibitions and dive headfirst into the squishy, squelchy joys of mud. The Boryeong Mud Festival is a reminder that sometimes, happiness is just a mud pit away.

Italy's Ivrea Orange Battle: A Fruity Fray

In a continuation of Italy's penchant for food-based combat, the Ivrea Orange Battle is a tradition stemming from the town's history of rebellion against tyranny. Participants, divided into teams, reenact a medieval insurrection by hurling oranges at each other with gusto. The streets turn into a citrusy battlefield, where the air is tangy with the zest of flying fruit. It's a vibrant, sticky, and somewhat perplexing spectacle that underscores the Italian spirit of community and resistance—albeit in a rather vitamin C-rich form.

As we close this chapter on the world's cultural curiosities, it's clear that the diversity of human celebration knows no bounds. From the serene to the explosive, the quirky to the downright muddy, these traditions remind us of the joy to be found in collective expression and the beauty of cultural diversity. They encourage us to look beyond our boundaries, to explore and appreciate the rich tapestry of global customs that make our world a more colorful, more interesting place. So, the next time you find yourself questioning a peculiar tradition, remember: it's these very oddities that make the tapestry of human culture so endlessly fascinating.

Chapter 8: Paranormal and Unexplained

Welcome, intrepid explorers of the odd and unexplained, to the twilight zone of our cultural curiosity tour. Here, we delve into the shadowy realms of the paranormal, where ghosts chat up the living, UFOs zip through our skies on intergalactic errands, and cryptids play hide-and-seek with eager cryptozoologists. So, grab your EMF meters, don your tinfoil hats, and let's explore the mysteries that make skeptics sigh and enthusiasts squeal with delight.

The Hauntingly Busy Afterlife of the Queen Mary

Docked permanently in Long Beach, California, the RMS Queen Mary is a retired ocean liner that's now more famous for its spectral passengers than its transatlantic crossings. Ghostly children laughing, phantom crew members still on duty, and the occasional spectral flapper from the 1930s are just part of the ship's non-living entourage. Tours offer the chance to meet these eternal travelers, providing a unique blend of history, luxury, and the kind of chills that don't come from the ocean breeze. It's as if the Queen Mary decided retirement was too boring without a few ghost stories.

Bigfoot: The Hide-and-Seek Champion of the World

Bigfoot, also known as Sasquatch, is arguably the world's most elusive celebrity. This hirsute hermit of the North American wilderness has been playing peek-a-boo with humans for centuries, leaving behind nothing but fuzzy photos, ambiguous footprints, and heated debates. Despite the lack of clear evidence, Bigfoot's fan base remains as dedicated as ever, proving that you don't need to be seen to be believed. Bigfoot continues to hold the title for the world's longest ongoing game of hide-and-seek, and frankly, we're all still eagerly playing along.

The Bermuda Triangle: The Ultimate Vanishing Act

The Bermuda Triangle, that infamous stretch of ocean between Miami, Bermuda, and Puerto Rico, has a rap sheet that includes disappearing ships, planes, and even the occasional sense of direction. While skeptics blame magnetic anomalies, bad weather, or good old human error, others prefer tales of alien abductions and time portals. The Triangle might just be the ocean's way of reminding us that, every now and then, it likes to keep a few secrets to itself—along with a collection of lost compasses and puzzled pilots.

The Loch Ness Monster: Scotland's Elusive Aquatic Star

Loch Ness, Scotland, is home to Nessie, the most camera-shy aquatic creature in folklore. Despite numerous sightings, grainy photographs, and sonar blips, this legendary lake resident insists on maintaining a mystique that keeps enthusiasts and skeptics alike coming back for more. Whether you believe Nessie is a prehistoric holdover or just a series of exceptionally buoyant logs, there's no denying the charm and tourism dollars this elusive monster brings to the Scottish Highlands. Nessie proves that in the world of cryptids, sometimes the mystery is more captivating than the monster.

CHAPTER 8: PARANORMAL AND UNEXPLAINED

The Haunted Dolls of Mexico's Island of the Dolls

Just south of Mexico City lies a small island with a big reputation for creepiness. Isla de las Muñecas (Island of the Dolls) is adorned with hundreds of dolls hanging from trees, placed there by the island's former caretaker in memory of a drowned girl. These eerie effigies, with their blank stares and silent whispers, have given the island an atmosphere that could make even the bravest ghost hunter think twice. It's a unique testament to the power of memory, belief, and the undeniable creep factor of dolls in large quantities.

Crop Circles: The Art Galleries of Aliens... or Pranksters?

Crop circles, those intricate designs mysteriously flattened into fields overnight, have puzzled onlookers since they first appeared. Are they messages from extraterrestrial visitors, natural phenomena, or just the handiwork of artists with a penchant for geometry and a love of secrecy? While many have been revealed as human-made, the debate continues, fueled by new formations each year. Crop circles remind us that, whether alien or artist, someone out there has a lot of time on their hands and a serious love for agrarian art.

The Spontaneous Human Combustion Mystery

Spontaneous Human Combustion (SHC) is the paranormal phenomenon where a person supposedly bursts into flames without an apparent external source of ignition. Though science points to more mundane explanations, the idea that one could suddenly go up in smoke during dinner keeps the mystery alive in popular culture. It's a chilling reminder to maybe skip the beans and opt for a salad, just in case.

The Philadelphia Experiment: A Naval Nautical Nonsense?

The Philadelphia Experiment allegedly was a secret U.S. Navy project in 1943 that made the USS Eldridge invisible to enemy devices. Depending on who you ask, it also teleported the ship from Philadelphia to Norfolk or caused crew members to fuse with the ship's hull. While firmly in the realm of science fiction, the story persists, a testament to our fascination with conspiracy theories and the limits of military science. It's like the Navy's version of a magic trick, where the rabbit disappears and leaves behind only questions.

Diving deeper into the twilight realm where the line between the explained and the utterly baffling becomes as murky as a haunted swamp at midnight, let's continue our jaunt through the world of the paranormal and the delightfully unexplained. These tales serve as a reminder that, no matter how much we learn, the universe always has a few tricks up its sleeve, ready to surprise us when we least expect it.

The Enigmatic Mothman of Point Pleasant

In the small town of Point Pleasant, West Virginia, a creature known as the Mothman became the center of a chilling local legend. Described as a towering figure with glowing red eyes and massive wings, the Mothman was spotted numerous times in the 1960s, becoming a harbinger of doom for many. Whether a misunderstood animal, an alien visitor, or the product of mass hysteria, the Mothman's legacy endures, inspiring books, movies, and a dedicated festival. It's as if Point Pleasant was chosen by lottery to be the backdrop of a live-action thriller, starring a creature that would make Batman think twice about his costume choices.

The Time-Slipping Village of Versailles

CHAPTER 8: PARANORMAL AND UNEXPLAINED

In 1901, two English women visiting the Palace of Versailles claimed to have slipped back in time to the French Revolution. They reported seeing buildings and scenes that no longer existed and even the ghost of Marie Antoinette herself. This temporal anomaly, often cited as evidence of time travel or dimensional slips, remains a classic tale of historical tourism taken to an unexpected level. It's a reminder that, sometimes, history is not content to stay in the past, and a good guidebook should probably include a section on temporal etiquette.

The Mysterious Disappearance of Flight 19

In 1945, Flight 19, a group of five U.S. Navy bombers, vanished off the coast of Florida while on a training flight. Despite extensive searches, neither the planes nor the crew were ever found, fueling speculation about alien abductions, interdimensional portals, and, of course, the Bermuda Triangle. The disappearance of Flight 19 remains one of aviation's most enduring mysteries, a sobering reminder of the ocean's vastness and the limits of our mastery over nature. It's as if the Atlantic decided to keep a few secrets of its own, tucked away beneath its waves.

The Taos Hum: Earth's Background Music?

In the small town of Taos, New Mexico, residents have reported hearing a persistent, low-frequency humming sound with no identifiable source. The "Taos Hum" has eluded explanation, leading to theories ranging from secret military projects to geological phenomena. It's the acoustic equivalent of a mystery novel that everyone's talking about but no one can find in the bookstore. The hum serves as a low-key reminder that the planet Earth might just have its soundtrack, albeit one that's not to everyone's taste.

The Fairy Circles of Namibia: Nature's Crop Circles

In the Namib Desert, there are thousands of nearly perfect circles of bare ground known as "fairy circles," which have puzzled scientists and tourists alike. These formations appear in the grasslands, with theories about their origins ranging from termite activity to plant competition for water. It's as if Mother Nature, in a whimsical mood, decided to dabble in landscape art, reminding us that she's still full of surprises and possibly harbors a secret passion for geometry.

The Phoenix Lights: A Celestial Traffic Jam?

In 1997, the skies over Phoenix, Arizona, were lit up by a series of widely witnessed, unexplained lights, forming what appeared to be a massive, stationary, V-shaped object. The "Phoenix Lights" remain one of the largest UFO sightings in history, with explanations ranging from military flares to extraterrestrial spacecraft. It's as though the universe decided to throw a light show, and Earth was the venue of choice, sparking debates and wonder long after the lights faded.

As we conclude this foray into the realms of the unexplained and the paranormal, it's clear that our world is far richer and stranger than we could ever imagine. From cryptids lurking in the shadows to historical anomalies that challenge our understanding of time, these mysteries invite us to look beyond the mundane and embrace the unknown. They remind us that curiosity is one of humanity's greatest traits, fueling our quest for knowledge and our love for a good mystery. So, keep your mind open, your flashlight close, and never stop exploring the endless possibilities that lie just beyond the edge of the explained. After all, the next great mystery is always waiting just around the corner, ready to be discovered, or perhaps, to discover us.

Chapter 9: Pioneers of the Unusual

Buckle up, dear readers, as we embark on a journey through the annals of history and modernity to shine a spotlight on those intrepid souls who dared to be different. These are the mavericks, the eccentrics, the square pegs in round holes – individuals who looked at the world and thought, "You know what this needs? More weird." From inventors of the impractical to artists who threw the rulebook into a blender, this chapter celebrates the pioneers of the unusual, those who remind us that sometimes, the road less traveled is paved with glitter and lined with rubber chickens.

Nikola Tesla: The Eccentric Genius

First up is Nikola Tesla, the inventor extraordinaire and undisputed champion of "Did he just do that?" Tesla's contributions to the development of electricity and magnetism are legendary, but it's his more eccentric projects that really capture the imagination. From attempting to develop a death ray to claiming he could split the Earth in two, Tesla's ambitions knew no bounds. He even fell in love with a pigeon, proving that genius often comes with a hefty side of quirkiness. Tesla reminds us that behind every textbook discovery, there might just be a mad scientist with a soft spot for birds.

The Countess of Lovelace and Her Mechanical Poetry

Ada Lovelace, often hailed as the world's first computer programmer, saw poetry in numbers and envisioned a future where machines could create art. In the 1840s, she worked on Charles Babbage's Analytical Engine and published notes that included what is essentially the first computer algorithm. Lovelace's vision of computational creativity was centuries ahead of its time, proving that sometimes, the line between poetry and programming is thinner than we think. She's the patron saint of digital artists everywhere, a reminder that code can be beautiful and that mathematics might just be the language of the universe.

Salvador Dalí: The Mustache with a Man Attached

No exploration of the unusual would be complete without a nod to Salvador Dalí, the mustachioed maestro of surrealism. With a flair for the dramatic and a penchant for the bizarre, Dalí turned every aspect of his life into a work of art, from his opulent home to his flamboyant public appearances. Whether he was walking an anteater on a leash or arriving at a lecture in a limousine filled with cauliflower, Dalí embodied the essence of eccentricity. He reminds us that the world is a canvas and that normalcy is vastly overrated.

Hetty Green: The Witch of Wall Street

Hetty Green, dubbed "The Witch of Wall Street," was a financial titan in a time when women were expected to be anything but. Known for her frugality and her formidable investment savvy, Green amassed a fortune in the late 19th and early 20th centuries. She conducted her business at the bank because she refused to rent an office and wore the same black dress to save money, all while outmaneuvering her male counterparts on Wall Street. Hetty Green proves that being unconventional isn't just a matter of personality but can be a strategic choice, especially when it comes to breaking glass ceilings with a well-aimed coin purse.

The Singular Style of Lord Timothy Dexter

Lord Timothy Dexter, an 18th-century American businessman, was known for his bizarre business decisions and even stranger lifestyle. He made a fortune by sending warming pans to the West Indies, a tropical climate where they were used for molasses and sugar processing, not bed-warming. Dexter also penned a book, "A Pickle for the Knowing Ones," which featured no punctuation and nonsensical prose. He even faked his own death to see how people would react. Dexter's life was a series of unorthodox choices that somehow paid off, making him a pioneer of the "it's so crazy, it just might work" philosophy.

The Whimsical World of Dr. Seuss

Theodor Seuss Geisel, known to the world as Dr. Seuss, turned children's literature on its head with his imaginative characters, playful rhymes, and whimsical worlds. His stories, filled with creatures like the Cat in the Hat and the Grinch, have delighted generations of readers. Dr. Seuss reminds us that imagination has no age limit, and that a little bit of silliness is a vital ingredient in the recipe of life.

Continuing our celebration of eccentricity and pioneering spirits, let's delve further into the annals of history and modern culture to spotlight a few more individuals whose unconventional paths have left indelible marks on the world. These characters remind us that sometimes, to make a difference, you've got to break the mold, dance to the beat of your own drum, and maybe even invent a new instrument while you're at it.

George Washington Carver: The Plant Wizard

George Washington Carver, often dubbed the "Plant Wizard," transformed agriculture in the American South. Born into slavery and later becoming a prominent scientist and inventor, Carver advocated for crop rotation and introduced alternative crops like peanuts and sweet potatoes to prevent soil depletion. But it wasn't just his agricultural insights that set him apart; Carver was an early environmentalist and a visionary who saw the potential in the humble peanut to create over 300 products, ranging from ink to soap to cooking oil. He showed the world that innovation doesn't always come from high-tech labs; sometimes, it grows right in your backyard.

Phineas Taylor Barnum: The Greatest Showman on Earth

P.T. Barnum, the man who coined the phrase "There's a sucker born every minute," was a master of spectacle and one of the founding figures of modern entertainment. Barnum's American Museum became a gathering place for the curious, showcasing everything from the Feejee mermaid to General Tom Thumb. Later, his traveling circus, "The Greatest Show on Earth," redefined public entertainment. Barnum's flair for the dramatic and his understanding of public curiosity turned hoaxes and hyperbole into an art form, reminding us that sometimes, what we seek is not the truth, but the thrill of the extraordinary.

The Audacious Escapades of Harry Houdini

Harry Houdini, born Erik Weisz, was the epitome of the escape artist, making a name for himself by wriggling out of handcuffs, straitjackets, and sealed water tanks. But Houdini's legacy isn't just about his death-defying stunts; it's also about his relentless pursuit of authenticity in the realm of the supernatural. A skeptic of mediums and spiritualists, Houdini dedicated part of his career to debunking fraudulent spiritualists, using his knowledge of illusion to expose their tricks. Houdini teaches us that true magic lies not in deceiving others, but in inspiring them with feats of skill, daring, and a dash of the impossible.

Marie Curie: The Radiant Scientist

Marie Curie, the first woman to win a Nobel Prize and the only person to win in two different scientific fields (Physics and Chemistry), was a pioneer of radioactivity—a term she coined herself. Curie's relentless research, often conducted under precarious conditions, led to the discovery of polonium and radium and opened new avenues for medical and scientific research. Her dedication to science, despite the societal constraints of her time and the personal toll of her research, showcases the extraordinary power of curiosity and resilience. Curie's life is a glowing testament to the idea that breaking barriers requires a blend of brilliance and stubbornness.

Hunter S. Thompson: The Gonzo Journalist

Hunter S. Thompson, the father of Gonzo journalism, threw objectivity out of the window and dove headfirst into his stories, becoming a central figure in his explosive, first-person narratives. From riding with the Hells Angels to running for sheriff on a "Freak Power" ticket, Thompson's life was as colorful and chaotic as his writing. He showed us that journalism could be a wild ride, blending fact with feverish imagination and proving that the truth is sometimes stranger—and more compelling—than fiction.

As we bring our journey through the lives of the unconventional to a close, it's clear that the fabric of history is woven with the threads of individuality and innovation. These pioneers of the unusual, with their daring deeds and unorthodox ways, remind us that progress often requires a departure from the norm and that the world is richer for their contributions. They inspire us to embrace our quirks, to pursue our passions with zeal, and to remember that making a difference often means daring to be different. So, here's to the trailblazers, the oddballs, and the visionaries—may we all find a bit of their spirit within us, urging us to explore, to question, and to color outside the lines in our quest to leave our mark on the world.

Conclusion

And there we have it, folks—the grand tour through the wacky, the weird, and the downright bewildering aspects of our world, wrapped up in a neat little package we've affectionately called "Curious Chronicles: A Journey Through Uncommon Knowledge and Strange Facts." From cheese-rolling daredevils to cities that decided to play hide-and-seek with civilization, we've journeyed together through the annals of the unusual, proving that reality often outstrips even the wildest fiction.

As we draw the curtains on this carnival of curiosities, let's take a moment to reflect on what we've learned. First and foremost, that the world is infinitely more interesting than your average high school textbook would have you believe. Who needs to memorize the periodic table when you can regale your friends with tales of the Great Emu War or the man who tried to invent a rain machine to control the weather? (Spoiler: It didn't end well.)

We've discovered that innovation isn't always about creating the next iPhone. Sometimes, it's about asking, "Can I make beer from 5,000-year-old yeast?" or "What if I turn my entire island into a doll sanctuary?" These stories remind us that creativity knows no bounds, especially when it's left to ferment in a barrel of "Why not?"

Our explorations have also shown us that the line between genius and madness is as thin as a Bigfoot sighting—blurry, contested, but utterly captivating. The pioneers of the unusual, from Nikola Tesla to the inventor of the Pet Rock, walk this tightrope with the grace of a drunken flamingo, reminding us that sometimes, you have to be a little crazy to change the world.

But perhaps the most important lesson of all is that curiosity is the key to an extraordinary life. It's the spark that ignites the imagination, fuels the pursuit of knowledge, and leads us down paths less traveled. In a world that often values conformity and the comfort of the familiar, daring to be curious is an act of rebellion, a declaration that we are not content to accept the world as it is but are eager to explore what it could be.

So, dear readers, as we bid adieu to this compendium of oddities, I encourage you to carry forward the spirit of curiosity that has guided our journey. Let the stories of cheese-rolling enthusiasts, ghostly inhabitants of ocean liners, and artists who see the world through a lens of surrealism inspire you to find the extraordinary in the everyday. Remember that the world is brimming with mysteries waiting to be uncovered, and it's up to us, the seekers of the strange and the stewards of the silly, to discover them.

In the end, "Curious Chronicles" isn't just a book; it's an invitation to look at the world with wonder, to question the status quo, and to laugh in the face of the inexplicable. Because, in a universe as vast and varied as ours, the greatest adventure is in the pursuit of knowledge, no matter how peculiar the path may be. Here's to the curious, the bold, and the slightly bonkers—may our thirst for the unknown never be quenched. Cheers!

Printed in Great Britain
by Amazon